AF269575

EARTH'S LANDFORMS

CAVES

by Lisa J. Amstutz

PEBBLE
a capstone imprint

Pebble Emerge is published by Pebble, an imprint of Capstone.
1710 Roe Crest Drive
North Mankato, Minnesota 56003
www.capstonepub.com

Library of Congress Cataloging-in-Publication Data is available on the Library of Congress website.
ISBN 978-1-9771-2459-3 (library binding)
ISBN 978-1-9771-2635-1 (paperback)
ISBN 978-1-9771-2502-6 (ebook pdf)

Summary: Simple text and colorful photos describe caves, how they form, where they are located, and what is found inside them.

Image Credits
Capstone Press/Karon Dubke, 20; Shutterstock: Albert Russ, 8, Balate Dorin, 11, DanielFreyr, 7, donghero, cover, eFesenko, 9, faak, 19, Ivan Kurmyshov, 5, Jan Kvita, 16, Ljiljana Jankovic, back cover, Peter_mi, 17, Piotr Krzeslak, 4, Sementer, 15, Vietnam Stock Images, 18, Wangkun Jia, 13

Editorial Credits
Editor: Alesha Sullivan; Designers: Elyse White and Bobbie Nuytten, Media Researcher: Kelly Garvin; Production Specialist: Tori Abraham

Printed and bound in China.
003322

Table of Contents

Words in **bold** are in the glossary.

WHAT IS A CAVE?

Have you ever been in a cave? A cave is an opening in the ground. It can also be an opening in the side of a hill. A cave is a type of landform. Landforms cover the earth.

It is dark inside a cave. The air is cool and **damp**. Some caves have large rooms. These are big enough to walk in. Some have long tunnels.

HOW ARE CAVES MADE?

Caves can be made in many ways. Some are made by water. Over time, water breaks down soft rock. It makes a hole. The hole gets bigger and bigger.

Some caves are made by **volcanoes**. Hot liquid called lava comes out of a volcano. A hole is made as the lava cools.

Large pieces of ice can make an ice cave. Melted snow and ice make a stream. Over time, the stream wears away the ice until a cave is made.

Ocean waves can make caves too.

The waves crash against the shore.

Water slowly wears away rocks.

Sometimes a cave is left behind.

ALL ABOUT CAVES

Many caves form in damp places. Water drips inside the cave. **Minerals** in the drops of water build up over thousands of years.

Stalactites form where water drips. Look up! They hang from the cave ceiling. **Stalagmites** grow where the water lands. Watch your step! They stick up from the floor.

A group of caves is called a system.
Mammoth Cave in the United States
is the world's longest cave system.
It covers more than 400 miles
(644 kilometers).

The world's deepest cave is in Asia.
It is more than 1 mile (1.6 km) deep.
Scientists reached the bottom in
2018. It took them a week to get there!

Mammoth Cave

WHAT LIVES IN CAVES?

Many animals live in caves. Fish swim in the pools of water. Sometimes bears live in caves during the winter. Bats hang from the ceilings.

PAST AND PRESENT

Long ago, people lived in caves. They painted on the walls. They drew animals, people, and handprints. Some cave drawings are thousands of years old!

Many caves tell a story. They teach us about the past. They show us how they were made. They tell us about the people who once lived there.

Today, many people like to visit caves all over the world. They can be fun to explore. Most visitors take a tour of caves led by a **guide**. Would you like to visit a cave?

Make Stalactites

Grow your own stalactites!

You'll Need:

- 2 jars
- water
- tray
- Epsom salt
- 8-inch (20-centimeter) piece of string
- 2 paper clips

Instructions:

1. Fill each jar with water and set them on the tray.

2. Add Epsom salt to each jar and stir. Keep adding salt and stirring until no more salt will dissolve.

3. Soak the piece of string in the salt water. Attach a paper clip to each end of the string.

4. Place one end of the string in each jar. The string should dip slightly in the center.

5. Watch your stalactites grow over the next few days.

Glossary

damp (DAMP)—slightly wet

guide (GIDE)—a person who points out the way

mineral (MIN-ur-uhl)—a solid found in nature

scientist (SYE-un-tist)—a person who studies the world around us

stalactite (stuh-LAK-tite)—a growth that hangs from the ceiling of a cave and was formed by dripping water

stalagmite (stuh-LAG-mite)—a growth that stands on the floor of a cave and was formed by drips of water from above

volcano (vol-KAY-noh)—an opening in Earth's surface that sometimes sends out hot lava, steam, and ash

Read More

Castro, Rachel. *Amazing Caves Around the World.*
North Mankato, MN: Capstone Press, 2019.

Kalman, Bobbie. *What Are Landforms?* New York:
Crabtree Publishing Company, 2018.

Siemens, Jared. Caves. New York: AV2 by Weigl, 2017.

Internet Sites

Easy Science for Kids: All About Caves
https://easyscienceforkids.com/all-about-caves/

Kiddle: Cave Facts for Kids
https://kids.kiddle.co/Cave

Science Kids: Fun Cave Facts for Kids
http://www.sciencekids.co.nz/sciencefacts/earth/caves.html

Index